Published by: AoPS Incorporated
                  15330 Avenue of Science
                  San Diego, CA 92128
                  info@BeastAcademy.com

ISBN: 978-1-934124-62-8

Beast Academy is a registered
trademark of AoPS Incorporated.

Written by Jason Batterson and Kyle Guillet
Illustrated by Erich Owen
Additional Illustrations by Paul Cox
Colored by Greta Selman

Visit the Beast Academy website at BeastAcademy.com.
Visit the Art of Problem Solving website at ArtofProblemSolving.com.
Printed in the United States of America.
2022 Printing.

Become a Math Beast!
For additional books,
printables, and more, visit

BeastAcademy.com

This is Guide 5B in a four-book series:

Contents:

Lizzie
"The BOOkwOrm"
can name every
dragOn species
On
Beast Island
(alphabetically)

Alex
"The Executive"
Plans tO run fOr
city cOmptrOller
when he's Old enOugh
fOr public Office

Winnie
"The Firecracker" Feisty!
GrOws 50 times
her Original size when angry!
(nOt really, but it's fun
to draw her that way)

GrOgg (me)
"The DenOminatOr"
        ^least commOn

ALter EgO:
FractiOn JacksOn!

Mr. Wriggles

kraken
shOp Teacher

Favorite pattern?
Arrrrrgyle

Favorite hOliday?
Arrrrrbor Day

Favorite element?
~~Arrrrrrgon~~
GOld

FiOna
Math Team cOach

DOnated her hair to
"Braids fOr Mermaids"
this summer

PrOfessOr GrOk
Math Lab
(full Of bOOby traps)

"calamitOus clod"

cOnstantly
captured by

Ms. Q.
Math Teacher

spends a lOt
Of time
with Mr. A.

R&G
campus Maintenance
Engineer(s?)

Let me ride
in their
gOlf cart
Once!

Sgt. ROte
Gym Teacher

can bench press
three times
his Own bOdyweight!
(4lbs.)

The Headmaster
How to use this book

Welcome to Beast Academy!

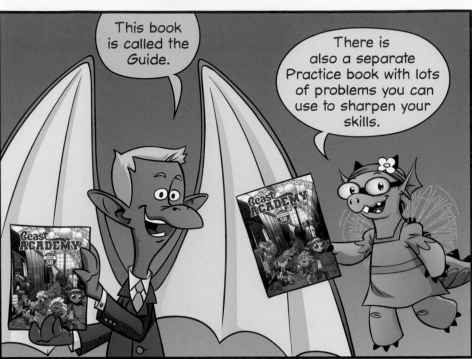

This book is called the Guide.

There is also a separate Practice book with lots of problems you can use to sharpen your skills.

The Guide is written like a comic book.

In a comic book, whatever I say shows up in these bubbles. They're called comic balloons.

Here's one!

Each character has a different balloon color. This makes it easy to tell who is talking.

My balloons are purple!

The story is told in panels.

Panels usually have a rectangular frame around them...

...like this one.

But sometimes, panels don't have frames...

...like the open panel I'm in now.

If you've gotten this far, you probably know a little bit about how to read a comic book.

You read a comic book the same way you read any other book... from left to right and from top to bottom.

On each page, start in the top left panel.

Go to the right, then down.

Read all of the balloons in each panel from left to right and top to bottom before moving to the next panel.

At first, you may need to think about which balloon to read next.

Like when lots of characters are talking.

Or when a character speaks more than once.

Right! And sometimes several balloons get connected.

With a little practice, reading comics becomes natural.

How many panels are on this page?

Contents: Chapter 4

See page 6 in the Practice book for a recommended reading/practice sequence for Chapter 4.

Chapter 4:
Statistics

Mediumest isn't a real word.

But, there is a name for the middle number in an ordered list.

The middle number is called the *median*.

The *median* weight of these nine rhinoceraptors is 1,986 pounds.

Name: Weight:
1. Herb 1,030 lbs
2. Jarvis 1,181 lbs
3. Bob 1,301 lbs
4. Curt 1,864 lbs
5. Hildi 1,986 lbs
6. Mort 2,098 lbs
7. Hildi 2,449 lbs

Cool.

Hildi has always seemed like a pretty ordinary rhinoceraptor.

Let's fill their feed bags.

The gals have already been fed. We need to fill the feed bags for the fellows.

MITZI

FEED

FEED

Here's how much feed each of the male rhinoceraptors gets.

Can you find the median amount of feed for these six male rhinoceraptors?

Name:	Feed:
Herb	76 lbs
Jarvis	90 lbs
Bob	93 lbs
Curt	119 lbs
Mort	123 lbs
Clive	164 lbs

Try it.

IN THIS BOOK, "AVERAGE" ALWAYS REFERS TO THE MEAN FOR A SET OF DATA.

Practice: Pages 7-18

*THE TOTAL WEIGHT OF ALL 100 PEARLS ABOVE IS $(99 \cdot 11) + (1 \cdot 3) = 1,089 + 3 = 1,092$ GRAMS. SO, THE AVERAGE WEIGHT OF THE 100 PEARLS IS $\frac{1092}{100} = 10\frac{92}{100} = 10.92$ GRAMS.

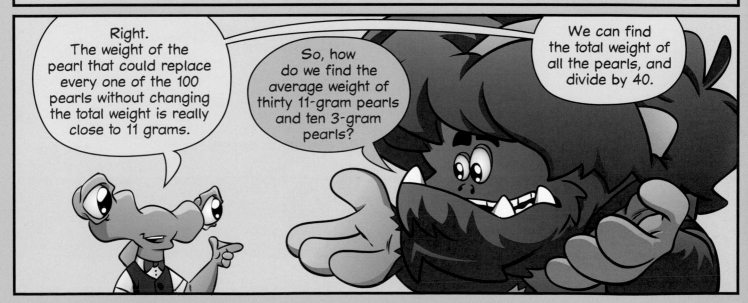

Each of the 30 pearls on this strand weighs 11 grams, so their total weight is 30 • 11 = 330 grams.

And each of the 10 pearls on this strand weighs 3 grams, for a total weight of 10 • 3 = 30 grams.

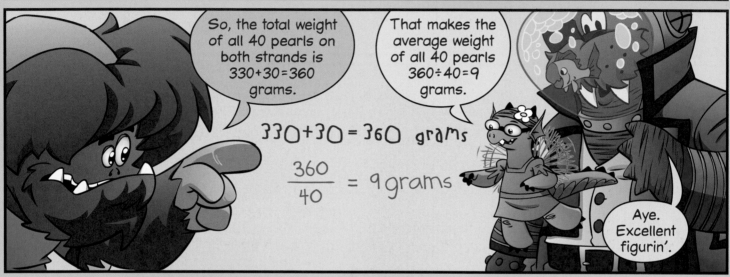

So, the total weight of all 40 pearls on both strands is 330 + 30 = 360 grams.

That makes the average weight of all 40 pearls 360 ÷ 40 = 9 grams.

$$330 + 30 = 360 \text{ grams}$$

$$\frac{360}{40} = 9 \text{ grams}$$

Aye. Excellent figurin'.

What's in here?

In this box be five gems...

...two identical rubies, 'n' three identical sapphires.

The average weight o' the five gemstones be 10 carats.

If the weight o' each ruby be 13 carats, what be the weight o' each sapphire?

A CARAT IS A MEASURE OF WEIGHT USUALLY USED TO MEASURE JEWELS. 1 CARAT = 200 MILLIGRAMS (0.2 GRAMS).

Try it.

So, each of the 3 sapphires has to be 6÷3=2 carats below the average.

That's 10−2=8 carats each.

13 +3
13 +3   +6
8   −2
8   −2
8   −2   −6

Aye. 'Tis a fine method.

Each o' the sapphires be 8 carats in weight.

Let's look at one more treasure.

In this satchel be a number o' coins.

One o' the coins be worth $50. The rest be worth $5 each.

The average value o' the coins be $8. How many coins be there in this satchel?

One coin be worth $50.
All others be $5 each.
The average value o' the coins be $8.

How many coins are in Captain Kraken's satchel?

31

We can write the total value of the coins two ways.

One coin be worth $50. All others be $5 each.

$$50+5n$$

...then the value of all the coins is $50+5n$ dollars.

If n is the number of $5 coins...

And since n is the number of $5 coins...

...there are $n+1$ coins all together.

If $n+1$ coins have an average value of $8, their total value is $8(n+1)$ dollars.

The average value o' the coins be $8.

$$50+5n = 8(n+1)$$

So, $50+5n = 8(n+1)$

$$50+5n = 8(n+1)$$
$$50+5n = 8n+8$$

To solve, we start by distributing the 8 to get $50+5n=8n+8$.

$$50+5n = 8(n+1)$$
$$50+5n = 8n+8$$
$$-5n \quad -5n$$
$$50 \quad = 3n+8$$

We subtract $5n$ from both sides so that we only have n on one side of the equation.

Then, we subtract 8 from both sides.

$$50+5n = 8(n+1)$$
$$50+5n = 8n+8$$
$$-5n \quad -5n$$
$$50 \quad = 3n+8$$
$$-8 \qquad -8$$
$$42 \quad = 3n$$

And we divide both sides by 3 to isolate n.

$n=14$.

$$50+5n = 8(n+1)$$
$$50+5n = 8n+8$$
$$-5n \quad -5n$$
$$50 \quad = 3n+8$$
$$-8 \qquad -8$$
$$\frac{42}{3} = \frac{3n}{3}$$
$$14 = n$$

Practice: Pages 19-23

What statistics have you learned about so far?

Statistics?

A *statistic* is a number used to describe a list of data.

Like the median, or the average?

Exactly.

We learned that the *median* is the middle number in a list of data.

And the *average* for a list of data is the number that could replace all of the numbers in a group without changing the group's sum.

STATISTICS IS THE FIELD OF MATHEMATICS INVOLVING THE STUDY OF DATA.

Great. What I'd like each of you to do is to create a list of seven numbers whose median is 8 and whose average is 11.

Try it.

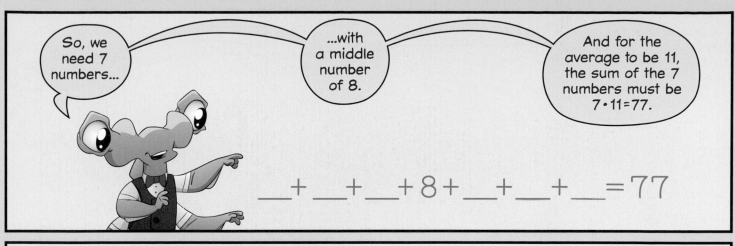

$$\_\_ + \_\_ + \_\_ + 8 + \_\_ + \_\_ + \_\_ = 77$$

Some of the lists are much more spread out than others.

My numbers only go from 8 to 15...

...but Grogg's start at 2 and go all the way up to 43!

8	8	8	8	15	15	15
2	2	2	8	10	10	43
2	4	6	8	11	18	28
3	3	3	8	20	20	20

We can label the difference between the largest and smallest value in each list.

Very good. The difference between the smallest and largest numbers in a list is called its *range*.

8	8	8	8	15	15	15		Range:
2	2	2	8	10	10	43		$15 - 8 = 7$
2	4	6	8	11	18	28		$43 - 2 = 41$
3	3	3	8	20	20	20		$28 - 2 = 26$
								$20 - 3 = 17$

We can list the most common number in each list.

The number that occurs most often in a list is called its *mode*.

8	8	8	8	15	15	15		Range:		Mode:
2	2	2	8	10	10	43		$15 - 8 = 7$		8
2	4	6	8	11	18	28		$43 - 2 = 41$		2
3	3	3	8	20	20	20		$28 - 2 = 26$		
								$20 - 3 = 17$		

But, what do we write if no number occurs more than once... ...or if there is a tie?

If no number in a list occurs more than once, we say that the list has no mode.

But, a list can have more than one mode if there is a tie. In Lizzie's list, there are two modes: 3 and 20.

Got it.

							Range:	Mode:
8	8	8	8	15	15	15	15 − 8 = 7	8
2	2	2	8	10	10	43	43 − 2 = 41	2
2	4	6	8	11	18	28	28 − 2 = 26	no mode
3	3	3	8	20	20	20	20 − 3 = 17	3 & 20

Good. You've named the four most common statistics used to describe a list of data: the median, average, range, and mode.

Median: Middle number

Average: $\dfrac{\text{Sum of numbers}}{\text{Number of numbers}}$

Range: Biggest minus smallest

Mode: Occurs most

Let's put these statistics to use.

The ages of the five monsters in a group are 11, 9, 10, 4, and 11 years.

Four new monsters join the group. This increases the average, median, and mode ages of the group by 1.

What is the new *range* of ages for the nine monsters in the group?

Ages: 11 9 10 4 11 ? ? ? ?

Try it.

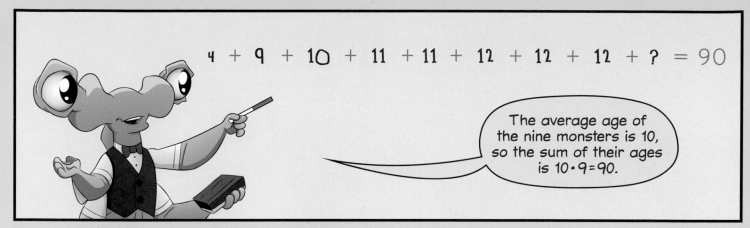

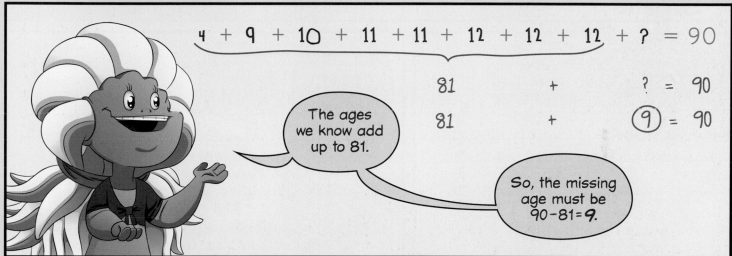

Practice: Pages 24-31

Stat Stumped is a game for 2 or more players played with a standard shuffled deck of cards with the face cards removed. Aces are 1's.

Setup:

Three cards are dealt onto the board to begin piles marked Mean, Median, and Range as shown. *If a 1 or a 10 is dealt onto the board, it is immediately moved below the board.*

Cards are then dealt so that there are a total of 7 cards below the board.

Stat!

Players simultaneously look for a group of three or more cards below the board that has at least two of the three statistics on the board.

A player who spots such a group calls out "Stat!" If the player can select a group of three or more cards from below the board that has at least two of the three statistics, the player gets to keep those cards in their score pile. Cards are then dealt from the stock so that there are 7 cards below the board, and play continues.

If the selected cards do *not* satisfy two of the three statistics, they are returned, and the other players are given the chance to call out "Stat!"

Stumped.

Any player who thinks it is impossible to create a group of cards with at least two of the stats says "Stumped."

Once all players are stumped, the first stumped player chooses one of the three stat piles and places the top card from the shuffled deck onto that pile. If the dealt card is an Ace, a 10, or matches the previous card, it is immediately moved below the board and play continues until someone calls "Stat!" or all players are stumped again.

Winning:

The game ends when all cards have been dealt from the stock and all players are stumped. The player who collects the most cards wins.

Stat Stumped

Samples:

Find a group of 3 or more cards below the board that has at least two of the three statistics on the board.

Answers are given at the bottom of this page.

1.
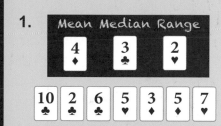

2.
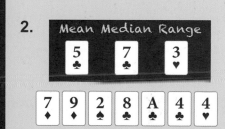

3.
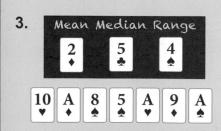

4.
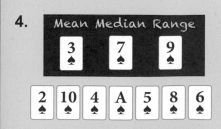

Find a partner and play!

Contents: Chapter 5

See page 32 in the Practice book for a recommended reading/practice sequence for Chapter 5.

Chapter 5: Factors & Multiples

Factors

132

What is the largest **odd** factor of 132?

We can list all the factors of 132.

Starting with 1, we know 1·132 = 132. So, we write 1·132 at the top of our list.

Next is 2·66.

Then 3·44.

132
1·132
2·66
3·44

REVIEW LISTING FACTORS IN CHAPTER 7 OF BEAST ACADEMY 4C.

4·33 =132.

5 is not a factor of 132, but 6 is. 6·22=132.

7, 8, 9, and 10 are *not* factors of 132. The last factor pair is 11·12=132.

132
1·132
2·66
3·44
4·33
6·22
11·12

To find the largest *odd* factor, we look for the biggest factor of 132 that is *not* divisible by 2.

Right. To find the largest odd factor of 132, we multiply all of the odd primes in its prime factorization.

3 · 11 = 33!

$$132 = 2 \cdot 2 \cdot ③ \cdot ⑪$$

33

Well done. A number's prime factorization tells us a lot about the number.

For example, look at the prime factorizations of these three numbers.

Which of these numbers is divisible by 12?

$$784 = 2^4 \cdot 7^2$$

$$858 = 2 \cdot 3 \cdot 11 \cdot 13$$

$$1,044 = 2^2 \cdot 3^2 \cdot 29$$

None of those numbers has 12 as a prime factor.

That's because 12 isn't *prime.*

But, we know that the prime factorization of 12 is 2 · 2 · 3.

How does that help?

Which number above is divisible by 12?

Any number that has at least two 2's and one 3 in its prime factorization is divisible by 12.

$$12 = 2^2 \cdot 3$$

I see. Since the prime factorization of 1,044 includes two 2's and a 3, $2 \cdot 2 \cdot 3 = 12$ is a factor of 1,044.

So, 1,044 is divisible by 12.

$$1{,}044 = 2^2 \cdot 3^2 \cdot 29$$
$$= (2 \cdot 2 \cdot 3) \cdot 3 \cdot 29$$
$$= (12) \cdot 3 \cdot 29$$

But if a number **doesn't** have two 2's and a 3 in its prime factorization...

...then it isn't divisible by 12.

784 doesn't have a 3 in its prime factorization...

...and the prime factorization of 858 only has one 2.

So, neither of those numbers is divisible by 12.

$$784 = 2^4 \cdot 7^2$$
$$858 = 2 \cdot 3 \cdot 11 \cdot 13$$

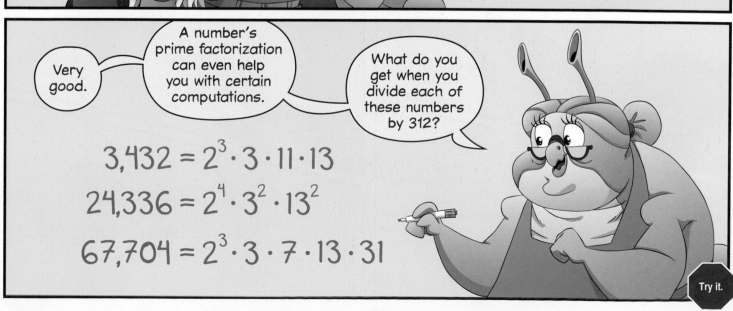

Very good.

A number's prime factorization can even help you with certain computations.

What do you get when you divide each of these numbers by 312?

$$3{,}432 = 2^3 \cdot 3 \cdot 11 \cdot 13$$
$$24{,}336 = 2^4 \cdot 3^2 \cdot 13^2$$
$$67{,}704 = 2^3 \cdot 3 \cdot 7 \cdot 13 \cdot 31$$

Try it.

To start, we can find the prime factorization of 312: $312 = 2^3 \cdot 3 \cdot 13$.

The prime factorization of 312 includes three 2's, a 3, and a 13.

Each of these three numbers includes three 2's, a 3, and a 13 in its prime factorization...

...so they are all divisible by 312.

If we separate three 2's, a 3, and a 13 from the rest of each prime factorization...

...we can write each number as a product of 312 and another number.

$$3{,}432 = 2^3 \cdot 3 \cdot 11 \cdot 13$$
$$24{,}336 = 2^4 \cdot 3^2 \cdot 13^2$$
$$67{,}704 = 2^3 \cdot 3 \cdot 7 \cdot 13 \cdot 31$$

$$3{,}432 = 2^3 \cdot 3 \cdot 11 \cdot 13$$
$$= (2 \cdot 2 \cdot 2 \cdot 3 \cdot 13) \cdot 11$$
$$= 312 \cdot 11$$

3,432 is $312 \cdot 11$.

So, $3{,}432 \div 312 = 11$.

$$3{,}432 \div 312 = 11$$

$$24{,}336 = 2^4 \cdot 3^2 \cdot 13^2$$
$$= 2 \cdot 2 \cdot 2 \cdot 2 \cdot 3 \cdot 3 \cdot 13 \cdot 13$$
$$= (2 \cdot 2 \cdot 2 \cdot 3 \cdot 13) \cdot (2 \cdot 3 \cdot 13)$$
$$= 312 \cdot 78$$

24,336 is $312 \cdot 78$.

So, $24{,}336 \div 312 = 78$.

$$24{,}336 \div 312 = 78$$

What is $67{,}704 \div 312$?

THE LAB
GCF

126   162

What is the largest number that is a factor of both 126 and 162?

We can list all of the factors of both numbers, and then find the biggest number that is in both lists.

126 has 12 factors.

162 has 10 factors.

The largest number that is a factor of both 126 and 162 is 18.

126

1 · 126
2 · 63
3 · 42
6 · 21
7 · (18)
9 · 14

162

1 · 162
2 · 81
3 · 54
6 · 27
9 · (18)

Well done. The largest factor that two numbers have in common is called their *greatest common factor.*

GCF, for short.

Try another pair. What is the GCF of 112 and 135?

112   135

GREATEST COMMON FACTOR (GCF) CAN ALSO BE CALLED GREATEST COMMON DIVISOR (GCD).

Try it.

50

We begin by looking at each number's prime factori--

YOINK!

Professor Grok?

Bwah Hah Hah! Professor Grok is gone! I've abducted your educator! It's time for something much more diabolically difficult!

The folly of finding the largest factor that divides two diminutive numerals is *undoubtedly dull*.

But finding the GCF of two four-digit numbers is a *formidable feat*.

The greatest integer that divides both 2,772 and 3,564 is located on the license plate of the livingmobile in the lower lot where I've locked your lecturer.

Liberate your professor lickety-split to *limit his laments*.

BEAST ACADEMY REUNION

2,772 & 3,564

Thunk

What is the GCF of 2,772 and 3,564?

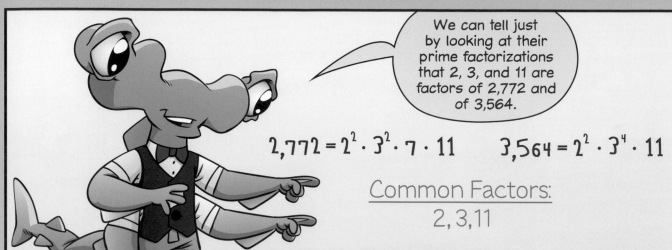

We can tell just by looking at their prime factorizations that 2, 3, and 11 are factors of 2,772 and of 3,564.

$$2{,}772 = 2^2 \cdot 3^2 \cdot 7 \cdot 11 \qquad 3{,}564 = 2^2 \cdot 3^4 \cdot 11$$

Common Factors:
2, 3, 11

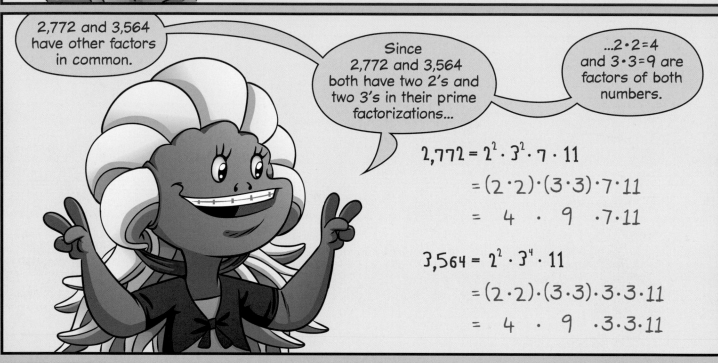

2,772 and 3,564 have other factors in common.

Since 2,772 and 3,564 both have two 2's and two 3's in their prime factorizations...

...2·2=4 and 3·3=9 are factors of both numbers.

$$2{,}772 = 2^2 \cdot 3^2 \cdot 7 \cdot 11$$
$$= (2 \cdot 2) \cdot (3 \cdot 3) \cdot 7 \cdot 11$$
$$= 4 \cdot 9 \cdot 7 \cdot 11$$

$$3{,}564 = 2^2 \cdot 3^4 \cdot 11$$
$$= (2 \cdot 2) \cdot (3 \cdot 3) \cdot 3 \cdot 3 \cdot 11$$
$$= 4 \cdot 9 \cdot 3 \cdot 3 \cdot 11$$

Since both numbers include a 2, a 3, and an 11 in their prime factorizations...

...both 2,772 and 3,564 also have 2·3·11=66 as a factor.

$$2{,}772 = 2^2 \cdot 3^2 \cdot 7 \cdot 11$$
$$= 2 \cdot 2 \cdot 3 \cdot 3 \cdot 7 \cdot 11$$
$$= (2 \cdot 3 \cdot 11) \cdot (2 \cdot 3 \cdot 7)$$
$$= 66 \cdot 42$$

$$3{,}564 = 2^2 \cdot 3^4 \cdot 11$$
$$= 2 \cdot 2 \cdot 3 \cdot 3 \cdot 3 \cdot 3 \cdot 11$$
$$= (2 \cdot 3 \cdot 11) \cdot (2 \cdot 3 \cdot 3 \cdot 3)$$
$$= 66 \cdot 54$$

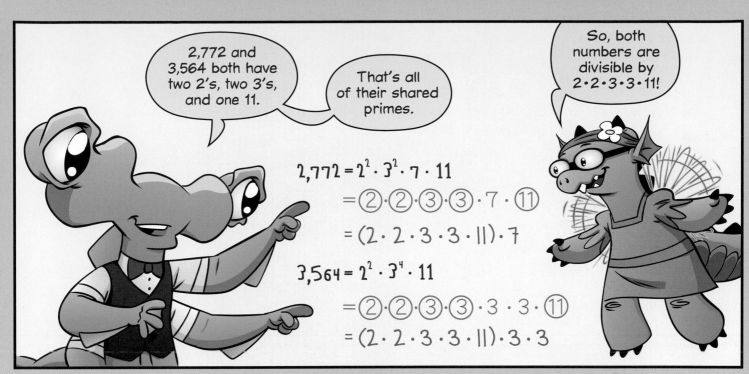

56

Practice: Pages 33-45

Setup

The game uses a standard deck of playing cards with the face cards and 10's (K, Q, J, 10) removed. Aces are treated as 1's. Players sit on opposite sides of a table. Shuffle and deal 2 cards to each player. Place the remaining cards face down in a pile called the stock. Flip two cards face-up on either side of the stock, as shown below. This creates two 2-digit numbers. In the example below, the player on one side of the table sees the numbers 42 and 15. The player on the opposite side of the table sees the numbers 51 and 24.

Player 1 Sees: 42   Stock   15

Player 2 sees: 24   Stock   51

Play

Players take turns placing one of the two cards in their hand on top of one of the four face-up cards. In the example below, Player 1 places the 4 of spades on the 5 of hearts. Now, Player 1 sees 42 and 14. Player 2 sees 41 and 24.
Each player finds the GCF of the pair of numbers they see, and adds the GCF to their score.

The GCF of 42 and 14 is 14, so Player 1 scores 14 points this turn. The GCF of 41 and 24 is 1, so Player 2 scores 1 point.

After playing a card and scoring, a player draws a card from the stock. This completes their turn.

Player 1 Sees: 42   Stock   14

Player 2 sees: 24   Stock   41

Winning

Players take turns playing a single card, scoring, and drawing until the stock is emptied and all the cards are played.
The player with the most points after all of the cards have been played wins.

Variations

For a simpler game, players are not given cards to begin, but instead take turns drawing from the stock and playing the card immediately. For a shorter game, choose a winning score. For example, the first player to go over 100 point wins.

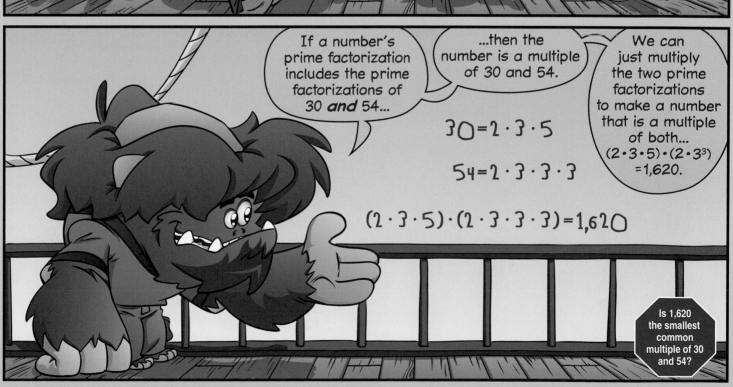

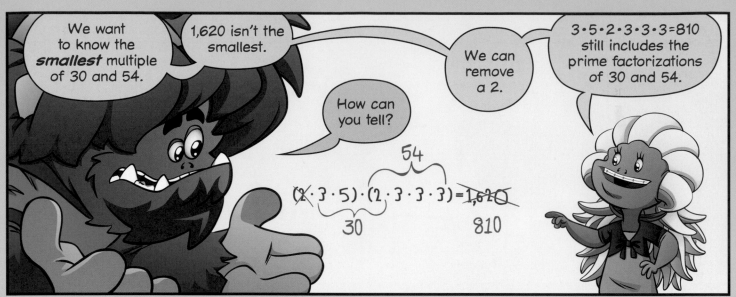

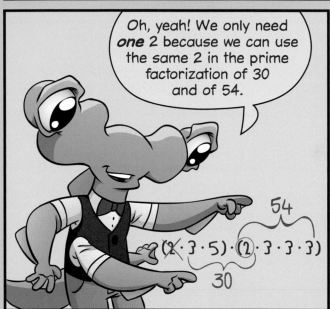

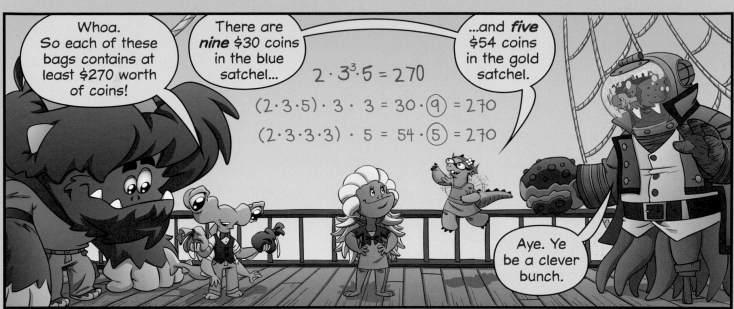

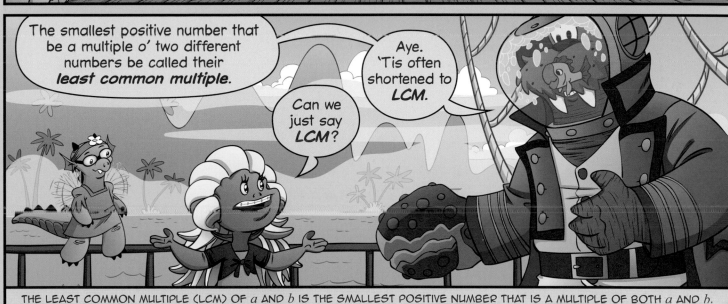

THE LEAST COMMON MULTIPLE (LCM) OF a AND b IS THE SMALLEST POSITIVE NUMBER THAT IS A MULTIPLE OF BOTH a AND b.

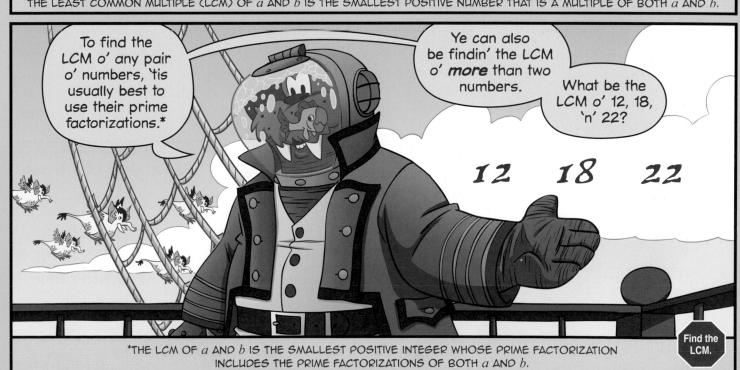

*THE LCM OF a AND b IS THE SMALLEST POSITIVE INTEGER WHOSE PRIME FACTORIZATION INCLUDES THE PRIME FACTORIZATIONS OF BOTH a AND b.

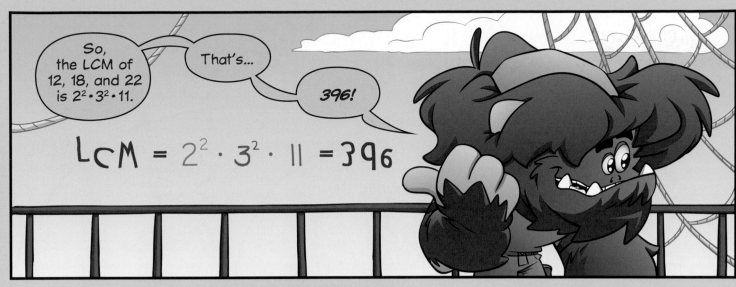

So, the LCM of 12, 18, and 22 is $2^2 \cdot 3^2 \cdot 11$.

That's...

396!

$$LCM = 2^2 \cdot 3^2 \cdot 11 = 396$$

Aye. The smallest number that be a multiple o' 12, 18, 'n' 22 be 396.

Not bad for a few junior buccaneers!

So, what are you planning to do with all of this loot, Captain Kraken?

Arrrr... I've been savin' up for a new basketball court on me deck.

65

Practice: Pages 46-55

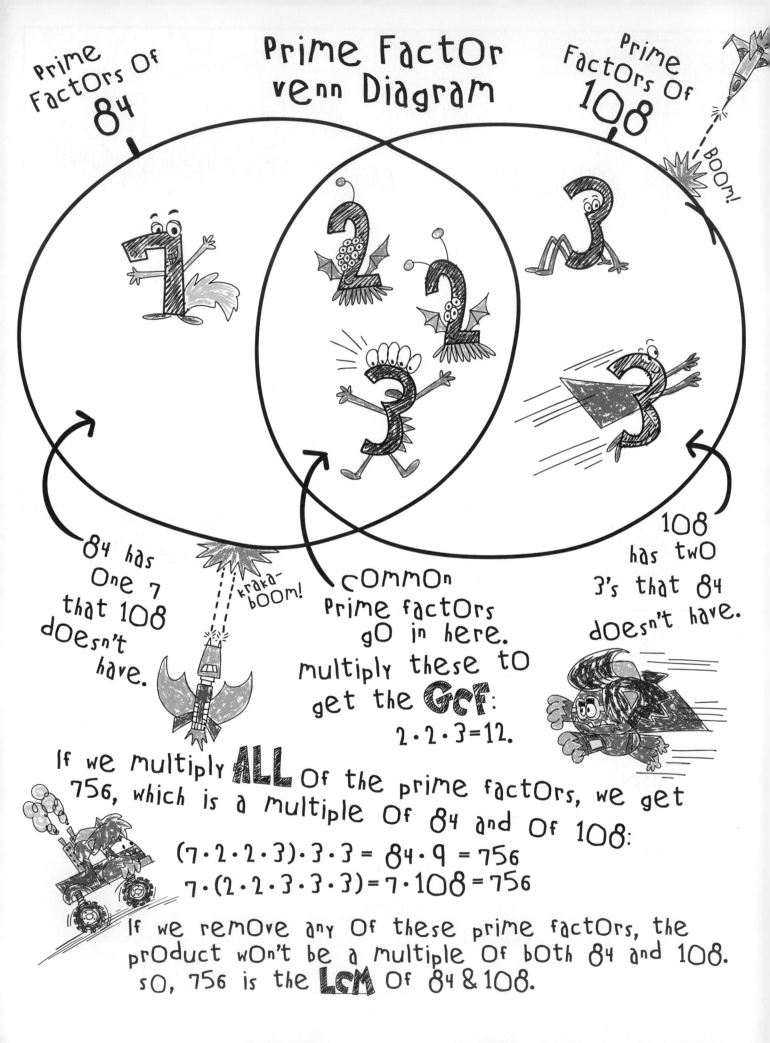

Prime Factors of 84

Prime Factors of 108

BOOM!

84 has One 7 that 108 doesn't have.

kraka-boom!

common prime factors go in here. multiply these to get the **GcF:** 2·2·3=12.

108 has two 3's that 84 doesn't have.

If we multiply **ALL** of the prime factors, we get 756, which is a multiple of 84 and of 108:

$$(7 \cdot 2 \cdot 2 \cdot 3) \cdot 3 \cdot 3 = 84 \cdot 9 = 756$$
$$7 \cdot (2 \cdot 2 \cdot 3 \cdot 3 \cdot 3) = 7 \cdot 108 = 756$$

If we remove any of these prime factors, the product won't be a multiple of both 84 and 108. so, 756 is the **LcM** of 84 & 108.

Today's problems involve *factorial* notation.

4!

Oh, right! I remember factorials.

FOUR!

The exclamation point does *not* mean that we shout the number.

Awwww.

WE FIRST LEARNED ABOUT FACTORIALS IN THE COUNTING CHAPTER OF BEAST ACADEMY 4B.

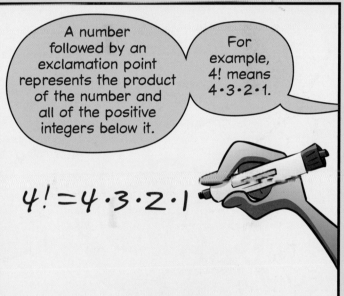

A number followed by an exclamation point represents the product of the number and all of the positive integers below it.

For example, 4! means 4·3·2·1.

$4! = 4 \cdot 3 \cdot 2 \cdot 1$

To begin, try writing the prime factorization of 8 factorial.

$8! = 8 \cdot 7 \cdot 6 \cdot 5 \cdot 4 \cdot 3 \cdot 2 \cdot 1$

Try it.

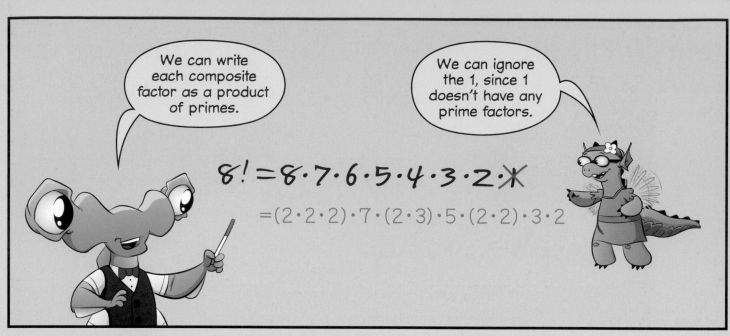

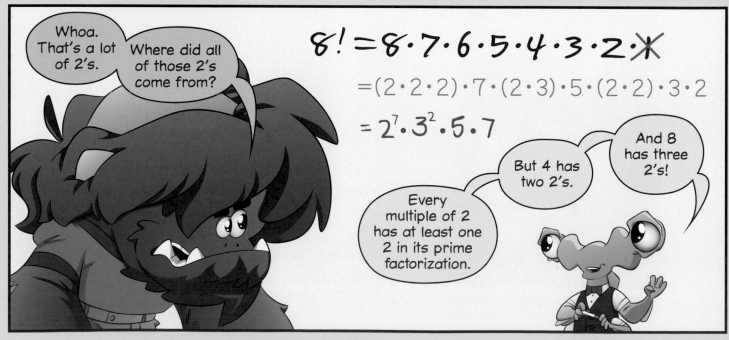

Good.

Let's take a look at a larger factorial.

How could we find the power of 3 in the prime factorization of 36 factorial?

$$36! = 36 \cdot 35 \cdot 34 \cdot \ldots \cdot 3 \cdot 2 \cdot 1$$

$$36! = 36 \cdot 35 \cdot 34 \cdot \ldots \cdot 3 \cdot 2 \cdot 1$$

Whoa. I do not want to write this prime factorization.

We don't need to find the **whole** prime factorization...

...just the power of 3.

$$36! = \boxed{36} \cdot 35 \cdot 34 \cdot \ldots \boxed{3} \cdot 2 \cdot 1$$

$$36 \cdot 33 \cdot 30 \cdot 27 \cdot 24 \cdot 21 \cdot 18 \cdot 15 \cdot 12 \cdot 9 \cdot 6 \cdot 3$$
$$\underline{3 \cdot 12} \quad \underline{3 \cdot 11} \quad \underline{3 \cdot 10} \quad \underline{3 \cdot 9} \quad \underline{3 \cdot 8} \quad \underline{3 \cdot 7} \quad \underline{3 \cdot 6} \quad \underline{3 \cdot 5} \quad \underline{3 \cdot 4} \quad \underline{3 \cdot 3} \quad \underline{3 \cdot 2} \quad \underline{3 \cdot 1}$$

12

Only the multiples of 3 have 3's in their prime factorizations.

We can ignore all of the other numbers and just focus on the multiples of 3.

From $3 \cdot 1 = 3$ to $3 \cdot 12 = 36$, there are twelve multiples of 3. So, there are at least twelve 3's in the prime factorization of $(36!)$.

Are there more 3's?

Are there?

THE NUMBER 500! HAS 1,135 DIGITS, INCLUDING ALL OF THE ZEROS AT THE END!

How many zeros are at the end of (500!)?

What kinds of numbers have zeros at the end?

Multiples of 10 end in at least one zero.

Multiples of 10 · 10 = 100 end in at least two zeros.

Multiples of 10 · 10 · 10 = 1,000 end in at least three zeros.

Every time you multiply by another 10, you get another zero.

$398 \cdot 10 = 3,980$

$19 \cdot 10 \cdot 10 = 1,900$

$643 \cdot 10 \cdot 10 \cdot 10 = 643,000$

THE ZEROS AT THE END OF A NUMBER ARE CALLED *TRAILING ZEROS*. FOR EXAMPLE, 643,000 HAS THREE TRAILING ZEROS.

So, we need to see how many 10's are in the prime factorization of 500 factorial.

Grogg! There are no 10's in the *prime* factorization of (500!).

10 is not prime.

Oh, right. Then, how can we find out how many 10's are in 500 factorial?

Try starting with an easier number, like 10 factorial.

How many zeros are at the end of 10 factorial?

Try it.

72

From 5·1=5 to 5·100=500, there are 100 multiples of 5.

So, there are at least 100 5's in the prime factorization of (500!).

Multiples of 5: 100

Multiples of 5·5=25 have at least two 5's in their prime factorizations.

So, each multiple of 25 adds another 5.

From 1·25=25 to 20·25=500, there are 20 multiples of 25.

That brings us to 100+20=120 5's in the prime factorization of (500!).

Multiples of 5: 100
Multiples of 25: 20

5·5·5=125 is the smallest number with three 5's in its prime factorization.

Multiples of 125 all have at least three 5's in their prime factorizations...

...125, 250, 375, and 500.

That's four more 5's.

Multiples of 5: 100
Multiples of 25: 20
Multiples of 125: 4

The smallest number with four 5's in its prime factorization is 5·5·5·5=625, so there are no more 5's to count.

There are 100+20+4=124 5's in the prime factorization of (500!).

We pair each 5 with a 2 to make a 10.

So, 500! ends in 124 zeros!

Multiples of 5: 100
Multiples of 25: 20
Multiples of 125: 4

Great work! You guys got that a lot faster than I did the first time I saw it.

I have some math questions I bet even *you* can't solve.

Let's hear them.

If there are 23 apples, and you take away 15, how many apples do you have?

23−15 is 8, so 8 apples are left...

...but, *I* have *15*.

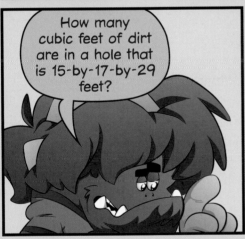

How many cubic feet of dirt are in a hole that is 15-by-17-by-29 feet?

Hmmm... 15·17·29 =7,395...

...but since it's a *hole*, there are *0* cubic feet of dirt in it.

OK, but you'll never get this one. I have two coins worth a total of 30 cents. One of the coins is not a nickel. What coins do I have?

There's only one way to make 30 cents with two coins.

A quarter and a nickel are worth 30 cents...

...and *one* of the coins is not a nickel.

Man, this guy is good!

75

Practice: Pages 56-63

Contents: Chapter 6

See page 64 in the Practice book for a recommended reading/practice sequence for Chapter 6.

Panel 1 (top):

We can do the same thing we did with the addition.

We can convert the fourths to twelfths, then subtract.

$\frac{3}{4} = \frac{9}{12}$.

$\frac{9}{12} - \frac{5}{12} = \frac{4}{12}$.

$$\frac{3}{4} - \frac{5}{12}$$
$$= \frac{9}{12} - \frac{5}{12}$$
$$= \frac{4}{12}$$
$$= \frac{1}{3}$$

And $\frac{4}{12}$ simplifies to $\frac{1}{3}$.

Panel 2 (middle):

Very good. To add or subtract fractions with different denominators, we can convert the fractions so that they have the same denominator.

This one is a little harder. How could you add $\frac{7}{10} + \frac{2}{15}$?

$$\frac{7}{10} + \frac{2}{15}$$

Panel 3 (bottom):

Well, we can't convert tenths to fifteenths.

Or fifteenths to tenths.

Maybe we can convert **both** fractions.

Huh?

81

$10 = 2 \cdot 5$

$15 = 3 \cdot 5$

LCM:

$2 \cdot 3 \cdot 5 = 30$

30 is the least common multiple of 10 and 15.

So, we can use 30 as the denominator of $\frac{7}{10}$ and $\frac{2}{15}$ to add them.

We can convert $\frac{7}{10}$ to $\frac{21}{30}$...

$$\overset{\times 3}{\underset{\times 3}{\frac{7}{10} = \frac{21}{30}}}$$

...and $\frac{2}{15}$ to $\frac{4}{30}$.

$$\overset{\times 2}{\underset{\times 2}{\frac{2}{15} = \frac{4}{30}}}$$

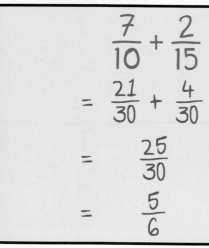

$$\frac{7}{10} + \frac{2}{15}$$
$$= \frac{21}{30} + \frac{4}{30}$$
$$= \frac{25}{30}$$
$$= \frac{5}{6}$$

So, $\frac{7}{10} + \frac{2}{15}$ equals $\frac{21}{30} + \frac{4}{30}$.

That's $\frac{25}{30}$, which simplifies to $\frac{5}{6}$.

$$\frac{3}{4} + \frac{4}{5} - \frac{5}{6} =$$

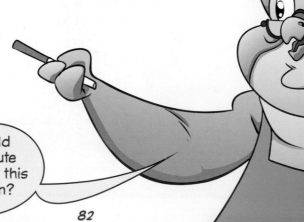

Nice work. Try one more, this time with three fractions.

How would you compute the value of this expression?

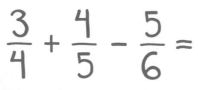

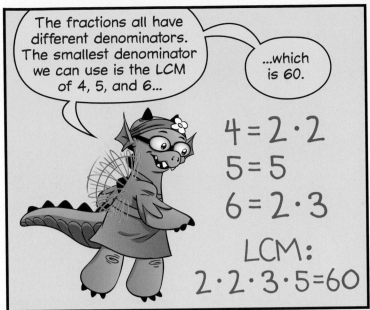

The fractions all have different denominators. The smallest denominator we can use is the LCM of 4, 5, and 6...

...which is 60.

$4 = 2 \cdot 2$
$5 = 5$
$6 = 2 \cdot 3$
LCM:
$2 \cdot 2 \cdot 3 \cdot 5 = 60$

$$\frac{3}{4} + \frac{4}{5} - \frac{5}{6} = \frac{45}{60} + \frac{48}{60} - \frac{50}{60}$$

So, we can convert all three fractions to sixtieths.

$$\frac{3}{4} + \frac{4}{5} - \frac{5}{6} = \frac{45}{60} + \frac{48}{60} - \frac{50}{60}$$
$$= \frac{45 + 48 - 50}{60}$$

We get 45+48−50 in the numerator...

...and 60 in the denominator.

$$\frac{3}{4} + \frac{4}{5} - \frac{5}{6} = \frac{45}{60} + \frac{48}{60} - \frac{50}{60}$$
$$= \frac{45 + 48 - 50}{60}$$
$$= \frac{43}{60}$$

So, it's $\frac{43}{60}$.

Excellent.

When do you little monsters leave for the World Math Olympiad Qualifiers?

Tomorrow morning.

We get to ride the train!

83

Practice: Pages 65-78

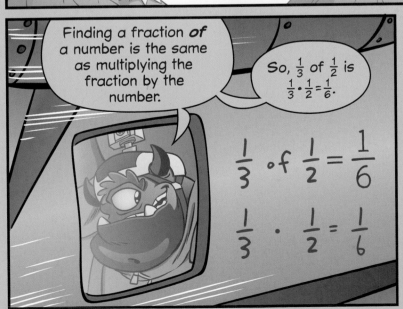

$\frac{1}{5} \cdot \frac{1}{9} = \frac{1}{5 \cdot 9}$, which is $\frac{1}{45}$.

Multiplying two unit fractions is as easy as multiplying their denominators.

$$\frac{1}{5} \cdot \frac{1}{9} = \frac{1}{45}$$

A FRACTION WITH 1 IN THE NUMERATOR IS CALLED A **UNIT FRACTION**.

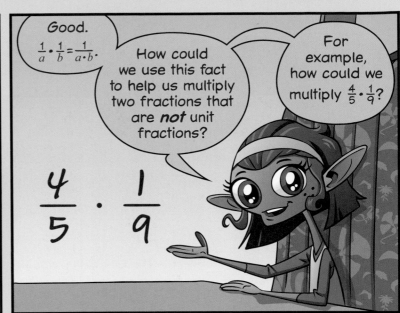

Good.
$\frac{1}{a} \cdot \frac{1}{b} = \frac{1}{a \cdot b}$.

How could we use this fact to help us multiply two fractions that are **not** unit fractions?

For example, how could we multiply $\frac{4}{5} \cdot \frac{1}{9}$?

$$\frac{4}{5} \cdot \frac{1}{9}$$

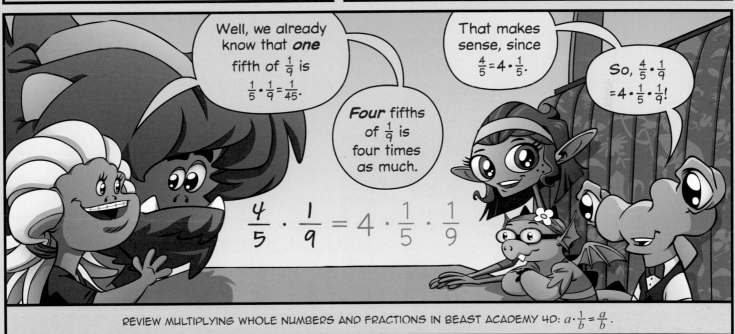

Well, we already know that **one** fifth of $\frac{1}{9}$ is $\frac{1}{5} \cdot \frac{1}{9} = \frac{1}{45}$.

Four fifths of $\frac{1}{9}$ is four times as much.

That makes sense, since $\frac{4}{5} = 4 \cdot \frac{1}{5}$.

So, $\frac{4}{5} \cdot \frac{1}{9} = 4 \cdot \frac{1}{5} \cdot \frac{1}{9}$!

$$\frac{4}{5} \cdot \frac{1}{9} = 4 \cdot \frac{1}{5} \cdot \frac{1}{9}$$

REVIEW MULTIPLYING WHOLE NUMBERS AND FRACTIONS IN BEAST ACADEMY 4D: $a \cdot \frac{1}{b} = \frac{a}{b}$.

$$\frac{4}{5} \cdot \frac{1}{9} = 4 \cdot \frac{1}{5} \cdot \frac{1}{9}$$
$$= 4 \cdot \frac{1}{45}$$
$$= \frac{4}{45}$$

So, $\frac{4}{5} \cdot \frac{1}{9} = \frac{4}{45}$.

Well done! Next, try $\frac{4}{5} \cdot \frac{7}{9}$.

$$\frac{4}{5} \cdot \frac{7}{9}$$

Try it.

89

Well, we just figured out that four fifths of **one** ninth is $\frac{4}{5} \cdot \frac{1}{9} = \frac{4}{45}$.

Four fifths of **seven** ninths must be seven times bigger.

Cool. We can write $\frac{4}{5}$ as $4 \cdot \frac{1}{5}$...

...and $\frac{7}{9}$ as $7 \cdot \frac{1}{9}$!

$$\frac{4}{5} \cdot \frac{7}{9} = \left(4 \cdot \frac{1}{5}\right) \cdot \left(7 \cdot \frac{1}{9}\right)$$

With just a little rearranging, we get $(4 \cdot 7) \cdot \left(\frac{1}{5} \cdot \frac{1}{9}\right)$...

$$\frac{4}{5} \cdot \frac{7}{9} = \left(4 \cdot \frac{1}{5}\right) \cdot \left(7 \cdot \frac{1}{9}\right)$$
$$= (4 \cdot 7) \cdot \left(\frac{1}{5} \cdot \frac{1}{9}\right)$$
$$= 28 \cdot \frac{1}{45}$$
$$= \frac{28}{45}$$

...which gives us $28 \cdot \frac{1}{45} = \frac{28}{45}$!

Wait...

...is that all we have to do?

What do you mean, Winnie?

To multiply $\frac{4}{5} \cdot \frac{7}{9}$, we just multiplied across the top and across the bottom.

$\frac{4}{5} \cdot \frac{7}{9} = \frac{4 \cdot 7}{5 \cdot 9}$!

Will that work for multiplying any two fractions?

$$\frac{4}{5} \cdot \frac{7}{9} \longrightarrow \frac{4 \cdot 7}{5 \cdot 9} = \frac{28}{45}$$

$$\frac{5}{16} \cdot \frac{24}{35} = \frac{5 \cdot 24}{16 \cdot 35} = \frac{24 \cdot 5}{16 \cdot 35} = \frac{24}{16} \cdot \frac{5}{35} = \frac{3}{2} \cdot \frac{1}{7} = \frac{3}{14}$$

...we can simplify $\frac{24}{16}$ to $\frac{3}{2}$, and $\frac{5}{35}$ to $\frac{1}{7}$.

That makes the multiplication way easier... $\frac{3}{2} \cdot \frac{1}{7} = \frac{3}{14}$!

Great job! How can simplifying first help you quickly compute this scary-looking product?

$$\frac{11}{12} \cdot \frac{12}{13} \cdot \frac{13}{14} \cdot \frac{14}{15} \cdot \frac{15}{16} \cdot \frac{16}{17} \cdot \frac{17}{18} \cdot \frac{18}{19}$$

We multiply all of the numerators and all of the denominators.

$$\frac{11}{12} \cdot \frac{12}{13} \cdot \frac{13}{14} \cdot \frac{14}{15} \cdot \frac{15}{16} \cdot \frac{16}{17} \cdot \frac{17}{18} \cdot \frac{18}{19}$$

$$= \frac{11 \cdot 12 \cdot 13 \cdot 14 \cdot 15 \cdot 16 \cdot 17 \cdot 18}{12 \cdot 13 \cdot 14 \cdot 15 \cdot 16 \cdot 17 \cdot 18 \cdot 19}$$

Then, we look for pairs of numbers that share a common factor.

$$\frac{11}{12} \cdot \frac{12}{13} \cdot \frac{13}{14} \cdot \frac{14}{15} \cdot \frac{15}{16} \cdot \frac{16}{17} \cdot \frac{17}{18} \cdot \frac{18}{19}$$

$$= \frac{11 \cdot 12 \cdot 13 \cdot 14 \cdot 15 \cdot 16 \cdot 17 \cdot 18}{12 \cdot 13 \cdot 14 \cdot 15 \cdot 16 \cdot 17 \cdot 18 \cdot 19}$$

Rearranging, we can get all of these pairs matched up.

$$= \frac{11 \cdot 12 \cdot 13 \cdot 14 \cdot 15 \cdot 16 \cdot 17 \cdot 18}{12 \cdot 13 \cdot 14 \cdot 15 \cdot 16 \cdot 17 \cdot 18 \cdot 19}$$

$$= \frac{12 \cdot 13 \cdot 14 \cdot 15 \cdot 16 \cdot 17 \cdot 18 \cdot 11}{12 \cdot 13 \cdot 14 \cdot 15 \cdot 16 \cdot 17 \cdot 18 \cdot 19}$$

$$\frac{11}{12} \cdot \frac{12}{13} \cdot \frac{13}{14} \cdot \frac{14}{15} \cdot \frac{15}{16} \cdot \frac{16}{17} \cdot \frac{17}{18} \cdot \frac{18}{19} = \frac{11 \cdot 12 \cdot 13 \cdot 14 \cdot 15 \cdot 16 \cdot 17 \cdot 18}{12 \cdot 13 \cdot 14 \cdot 15 \cdot 16 \cdot 17 \cdot 18 \cdot 19}$$

$$= \frac{12 \cdot 13 \cdot 14 \cdot 15 \cdot 16 \cdot 17 \cdot 18 \cdot 11}{12 \cdot 13 \cdot 14 \cdot 15 \cdot 16 \cdot 17 \cdot 18 \cdot 19}$$

$$= \frac{12}{12} \cdot \frac{13}{13} \cdot \frac{14}{14} \cdot \frac{15}{15} \cdot \frac{16}{16} \cdot \frac{17}{17} \cdot \frac{18}{18} \cdot \frac{11}{19}$$

$$= 1 \cdot 1 \cdot 1 \cdot 1 \cdot 1 \cdot 1 \cdot 1 \cdot \frac{11}{19}$$

$$= \frac{11}{19}$$

We end up with a bunch of 1's times $\frac{11}{19}$, which is just $\frac{11}{19}$!

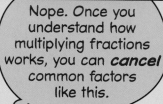

Nope. Once you understand how multiplying fractions works, you can *cancel* common factors like this.

We don't need to show all of that work, do we?

$$\frac{\overset{2}{\cancel{10}}}{11} \cdot \frac{8}{\underset{3}{\cancel{15}}} = \frac{16}{33}$$

Here, we cancel a common factor of 5, leaving a 2 in the numerator and a 3 in the denominator.

$$\frac{\overset{1}{\cancel{4}}}{7} \cdot \frac{3}{\underset{1}{\cancel{4}}} = \frac{3}{7}$$

Here, we cancel a common factor of 4.

WE DON'T **NEED** TO WRITE THE 1'S.

This is Max's stop.

There he is!

Hey, guys! Meet me in the lounge car.

94

Practice: Pages 79-88

Simplifying Fractions: Cancelling

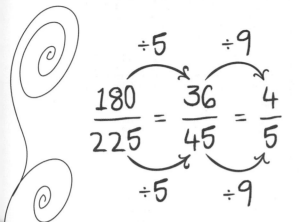

$$\frac{180}{225} = \frac{36}{45} = \frac{4}{5}$$

with $\div 5$ and $\div 9$ arrows above, $\div 5$ and $\div 9$ below

When we show the work like this, it is called "cancelling"

$$\frac{\overset{4}{\cancel{\overset{36}{\cancel{180}}}}}{\underset{5}{\underset{45}{\cancel{225}}}} = \frac{4}{5}$$

Cancelling factors in products:

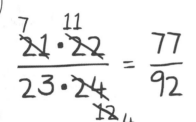

$$\frac{\overset{7}{\cancel{21}} \cdot \overset{11}{\cancel{22}}}{23 \cdot \underset{\underset{4}{12}}{\cancel{24}}} = \frac{77}{92}$$

We cancel a factor of 2 from 22 and 24.
We write 11 by the 22 and 12 by the 24.
Then, we cancel a factor of 3 from 21 and 12.
We write 7 by the 21 and 4 by the 12.

Cancelling when multiplying fractions:

$$\frac{\overset{3}{\cancel{9}}}{\underset{2}{\cancel{14}}} \cdot \frac{\overset{1}{\cancel{7}}}{\underset{4}{\cancel{12}}} = \frac{3}{8}$$

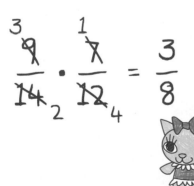

We cancel a factor of 3 from 9 and 12.
We write 3 by the 9 and 4 by the 12.
Then, we cancel a factor of 7 from 7 and 14.
We write 1 by the 7 and 2 by the 14.

$$\frac{\overset{3}{\cancel{6}}}{11} \cdot \frac{\cancel{3}}{\underset{5}{\cancel{10}}} \cdot \frac{7}{\underset{\cancel{3}}{\cancel{9}}} = \frac{7}{55}$$

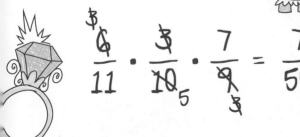

We cancel a factor of 2 from 6 and 10.
We cancel a factor of 3 from 3 and 9.
Then, we cancel the 3 on top and bottom.
We don't need to write 1's by the 3's.

Whoa... This train is cool!

Hi, guys, over here!

What are you working on?

I'm reviewing test questions from past competitions.

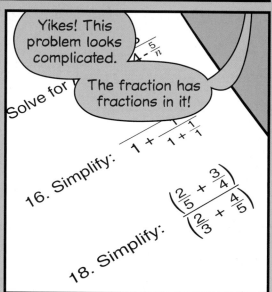

Yikes! This problem looks complicated.

The fraction has fractions in it!

Solve for

$$\dfrac{2 - \frac{5}{\pi}}{1 + \dfrac{1}{1 + \frac{1}{1}}}$$

16. Simplify:

18. Simplify: $\left(\dfrac{\frac{2}{5} + \frac{3}{4}}{\frac{2}{3} + \frac{4}{5}}\right)$

We know how to add fractions.

We can start by simplifying the numerator and denominator.

In the numerator we get $\frac{8}{20} + \frac{15}{20} = \frac{23}{20}$.

And in the denominator, we get $\frac{10}{15} + \frac{12}{15} = \frac{22}{15}$.

Now what do we do?

$$\left(\dfrac{\frac{2}{5} + \frac{3}{4}}{\frac{2}{3} + \frac{4}{5}}\right) = \left(\dfrac{\frac{8}{20} + \frac{15}{20}}{\frac{10}{15} + \frac{12}{15}}\right) = \dfrac{\left(\frac{23}{20}\right)}{\left(\frac{22}{15}\right)}$$

THIS IS VERY IMPORTANT, SO WE WILL REPEAT IT: TO DIVIDE BY A NUMBER, WE MULTIPLY BY ITS RECIPROCAL.

RECIPROCALS AND FRACTION DIVISION ARE INTRODUCED IN CHAPTER 10 OF BEAST ACADEMY 4D.

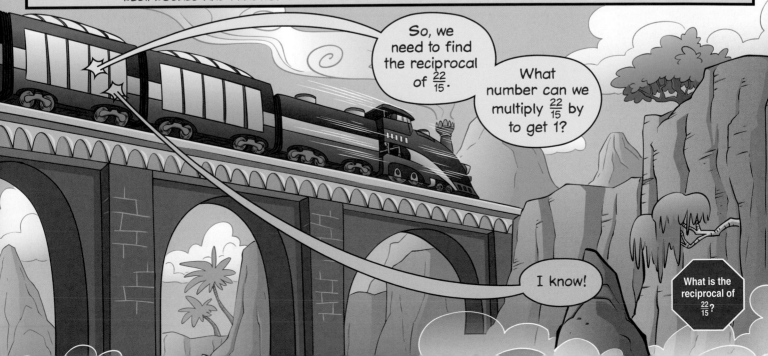

First, we can convert the mixed numbers to fractions.

Then, we change the division to multiplication.

To divide by $\frac{29}{6}$, we multiply by $\frac{6}{29}$.

$$5\frac{4}{5} \div 4\frac{5}{6} = \frac{29}{5} \div \frac{29}{6} = \frac{29}{5} \cdot \frac{6}{29}$$

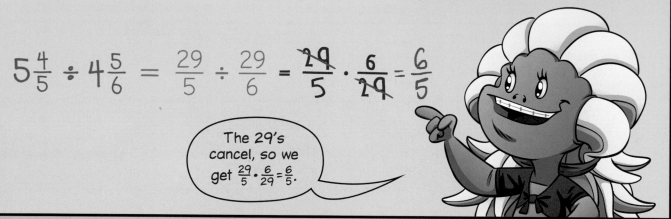

$$5\frac{4}{5} \div 4\frac{5}{6} = \frac{29}{5} \div \frac{29}{6} = \frac{\cancel{29}}{5} \cdot \frac{6}{\cancel{29}} = \frac{6}{5}$$

The 29's cancel, so we get $\frac{29}{5} \cdot \frac{6}{29} = \frac{6}{5}$.

So, it takes Sally Slug $\frac{6}{5} = 1\frac{1}{5}$ hours to slime $5\frac{4}{5}$ meters.

Nice. You guys learn fast!

Alright, team. We're almost there. Grab your stuff and let's get ready to go.

SKREE

MATH TEAM
The Final Round

Welcome to the final round of this year's World Math Olympiad Qualifiers for the Beast Island region.

I will now announce the top four teams. When I announce your team, please come to the stage and stand in your designated area.

Monsters, please welcome to the stage...

...the teams from Orb Elementary...

...Midvale School for the Gifted...

...Faun Crest Primary School...

...and Beast Academy.

Teams, please test your buzzers.

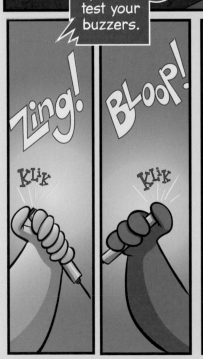

Zing!

Klik

Bloop!

Klik

Zerp!

Klik

Fwonk!

Klik

Perfect. Let's get to the first question.

How many seconds are in $\frac{1}{2}$ of $\frac{2}{3}$ of $\frac{3}{4}$ of $\frac{4}{5}$ of $\frac{5}{6}$ of one hour?

Fwonk!

How many?

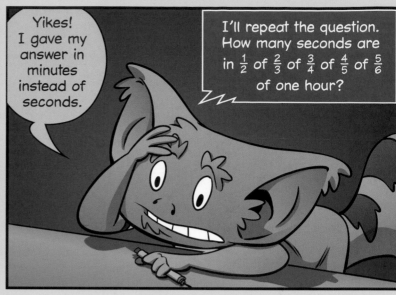

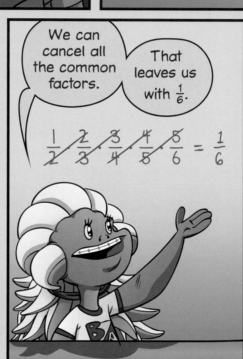

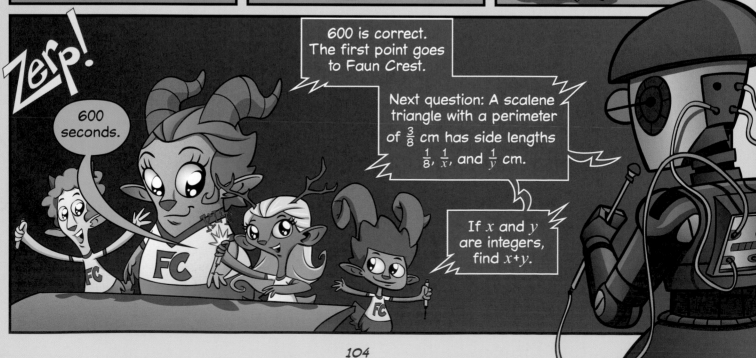

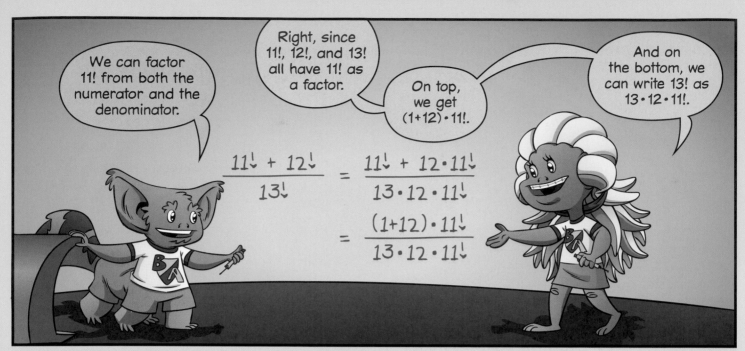

FIND DETAILED SOLUTIONS TO ALL FOUR PROBLEMS ABOVE AT BEASTACADEMY.COM.

Practice: Pages 89-97

Index

Want more Beast Academy?
Try Beast Academy Online!

Learn more at BeastAcademy.com